QuickStudy

PERCENT APPLICATIONS

% INCREASE

FORMULAS: $\frac{\%\text{ increase}}{100} = \frac{\text{amount of increase}}{\text{original value}}$ or

(original value) x (% increase) = amount of increase

If the amount of increase is not given, it may be found through this subtraction: **(new value) - (original value) = amount of increase.** *EXAMPLE:* The Smyth Company had 10,000 employees in 1992 and 12,000 in 1993. Find the % increase.

Amount of increase = 12,000 - 10,000 = 2,000

% increase: $\frac{n}{100} = \frac{2000}{10000}$ So n = 20 and the % increase = 20% because % means "out of 100."

% COMMISSION

FORMULAS: $\frac{\%\text{ commission}}{100} = \frac{\$\text{ commission}}{\$\text{ sales}}$

($ sales) x (% commission) = $ commission

EXAMPLE: Missy earned 4% on a house she sold for $125,000. Find her dollar commission.

% commission: $\frac{4}{100} = \frac{\$\text{commission}}{\$125,000}$

or ($125,000) x (4%) = $ commission, so $ commission = $5000.

% DECREASE

FORMULAS: $\frac{\%\text{ decrease}}{100} = \frac{\text{amount of decrease}}{\text{original value}}$ or

(original value) x (% decrease) = amount of decrease

If not given, amount of decrease = (original price) - (new value)

EXAMPLE: The Smyth Company had 12000 employees in 1993 and 9000 in 1994. Find the percent decrease.

Amount of decrease = 12000 - 9000 = 3000

% decrease: $\frac{n}{100} = \frac{3000}{12000}$

so n = 25 and the % discount = 25%

% DISCOUNT

FORMULAS: $\frac{\%\text{ discount}}{100} = \frac{\text{amount of discount}}{\text{original price}}$ or

(original price) x (% discount) = $ discount

If not given, ($ discount) = (original price) - (new price)

EXAMPLE: The Smyth Company put suits that usually sell for $250 on sale for $150. Find the percent discount.

% discount: $\frac{n}{100} = \frac{\$100}{\$250}$

so n = 40 and the % discount = 40%

% MARKUP

FORMULAS: $\frac{\%\text{ markup}}{100} = \frac{\$\text{ markup}}{\text{original price}}$ or

(original price) x (% markup) = $ markup

If not given, ($ markup) = (new price) - (original price)

EXAMPLE: The Smyth Company bought blouses for $20 each and sold them for $44 each. Find the percent markup.

$ markup = $44 - $20 = $24 % markup: $\frac{n}{100} = \frac{24}{20}$

so n = 120 and the % markup = 120%

% EXPENSES OR COSTS

FORMULAS: $\frac{\%\text{ expenses}}{100} = \frac{\$\text{ expenses}}{\text{total \$ income}}$

or **(total $ income) x (% expenses) = $ expenses**

EXAMPLE: The Smyth Company had a total income of $250,000 and $7500 profit last month. Find the percent expenses. $ expenses: = $250,000 - $7,500 = $242,500

% expenses: $\frac{n}{100} = \frac{242500}{250000}$

so n = 97 and the % expenses = 97%.

SIMPLE INTEREST

FORMULAS: $i = prt.$

or (total amount) = (principal) + interest

Where i = interest

p = principal; money borrowed or lent

r = rate; percent rate

t = time; expressed in the same period as the rate, i.e., if rate is per year, then time is in years or part of a year. If rate is per month, then time is in months.

EXAMPLE: Carolyn borrowed $5000 from the bank at 6% simple interest per year. If she borrowed the money for only 3 months, find the total amount that she paid the bank.

$ interest = prt = ($5000)(6%)(.25) = $75

Notice that the 3 months was changed to .25 of a year.

Total Amount = $p + i$ = $5000 + $75 = $5075

% PROFIT

FORMULAS: $\frac{\%\text{ profit}}{100} = \frac{\$\text{ profit}}{\text{total \$ income}}$

or (total $ income) x (% profit) = $ profit

if not given, $ profit = (total $ income) - ($ expenses).

EXAMPLE: The Smyth Company had expenses of $150,000 and a profit of $10,000. Find the % profit.

total $ income = $150,000 + $10,000 = $160,000

% profit: $\frac{n}{100} = \frac{10000}{160000}$

or ($160000) x ($n$) = $10000. In either case the % profit = 6.25%.

COMPOUND INTEREST

FORMULA: $A = p(1+\frac{r}{n})^{nt}$

Where: A = total amount

p = principal; money saved or invested

r = rate of interest; usually a % per year

t = time; expressed in years

n = total number of periods

EXAMPLE: John put $100 into a savings account at 4% compounded quarterly for 8 years. How much was in the account at the end of 8 years?

$A = p(1+\frac{r}{n})^{nt}$

$A = 100(1+\frac{.04}{4})^{(4x8)}$

$A = 100(1.01)^{32}$

$A = 100(1.3749)$

$A = 137.49$

"IS" AND "OF"

Any problems that are or can be stated with percent and the words "**is**" and "**of**" can be solved using these formulas:

FORMULAS: $\frac{\%}{100} = \frac{\text{"is" number}}{\text{"of" number}}$

or "**of**" means multiply and "**is**" means equals.

EXAMPLE 1: What percent **of** 125 **is** 50? $\frac{n}{100} = \frac{50}{125}$

or n x 125 = 50, in either case the percent = 40%.

EXAMPLE 2: What number is 125% of 80? $\frac{125}{100} = \frac{n}{80}$

or (1.25) (80) = n. In either case the number = 100.

ALGEBRA

VOCABULARY

- **Variables** are letters used to represent numbers.
- **Constants** are specific numbers that are not multiplied by any variables.
- **Coefficients** are numbers that are multiplied by one or more variables. *EXAMPLES:* -4xy has a coefficient of -4; $9m^3$ has a coefficient of 9; x has an invisible coefficient of 1.
- **Terms** are constants or variable expressions. *EXAMPLES:* 3a; $-5c^4d$; $25mp^3r^5$; 7 are all terms.
- **Like or similar terms** are terms that have the same variables to the same degree or exponent value. Coefficients do not matter, they may be equal or not. *EXAMPLES:* $3m^2$ and $7m^2$ are like terms because they both have the same variable to the same power or exponent value. - $15a^6b$ and $6a^6b$ are like terms, but $2x^4$ and $6x^3$ are not like terms because although they have the same variable, x, it is to the power of 4 in one term and to the power of 3 in the other.
- **Algebraic expressions** are terms that are connected by either addition or subtraction. *EXAMPLES:* $2s + 4a^2 - 5$ is an algebraic expression with 3 terms, 2s and $4a^2$ and 5.
- **Algebraic equations** are statements of equality between at least two terms. *EXAMPLES:* 4z = 28 is an algebraic equation. 3(a - 4) + 6a = 10 - a is also an algebraic equation. Notice that both statements have equal signs in them.
- **Algebraic inequalities** are statements that have either > or < between are least two terms. *EXAMPLES:* 50 < -2x is an algebraic inequality. 3 (2n + 7) > -10 is an algebraic inequality.

DISTRIBUTIVE PROPERTY FOR POLYNOMIALS

- Type 1: **a(c + d) = ac + ad**; *EXAMPLE:* $4x^3(2xy + y^2) = 8x^4y + 4x^3y^2$
- Type 2: **(a + b)(c+d) = a(c+d) + b(c+d) = ac + ad + bc + bd** *EXAMPLE:* $(2x + y)(3x - 5y) = 2x(3x - 5y) + y(3x - 5y) = 6x^2 - 10xy + 3xy - 5y^2 = 6x^2 - 7xy - 5y^2$. This may also be done by using the **FOIL Method for Products of Binomials** (See *ALGEBRA 1* chart). This is a popular method for multiplying 2 terms by 2 terms **only**. FOIL means first term times first term, outer term times outer term, inner term times inner term, and last term times last term.

COMBINING LIKE TERMS (ADDING OR SUBTRACTING)

- **RULE:** Combine (add or subtract) only the coefficients of like terms and never change the exponents during addition or subtraction. **a + a = 2a**

EXAMPLES: $4xy^3$ and $-7y^3x$ are like terms, even though the x and y^3 are not in the same order, and may be combined in this manner $4xy^3 - 7y^3x = -3xy^3$, notice only the coefficients were combined and no exponent changed; $-15a^2bc$ and $3bca^2$ are not like terms because the exponents of the a are not the same in both terms, so they may not be added or subtracted.

MULTIPLYING TERMS

- ***Definition:*** $3^5 = (3)(3)(3)(3)(3)$; that is, 3 is called the **base** and it is multiplied by itself 5 times because the **exponent** is 5. $a^m = (a)(a)(a)...(a)$; that is, the **a** is multiplied by itself **m** times.
- **Product Rule for Exponents:** $(a^m)(a^n) = a^{m+n}$; that is, when multiplying the same base, **a** in this case, simply add the exponents.
- Any terms may be multiplied, not just like terms.
- **RULE:** Multiply the coefficients and multiply the variables (this means you have to add the exponents of the same variable).

EXAMPLE: $(4a^4c)(-12a^2b^3c) = -48a^6b^3c^2$. Notice that 4 times -12 became -48, a^4 times a^2 became a^6, c times c became c^2, and the b^3 was written to indicate multiplication by **b**, but the exponent did not change on the **b** because there was only one **b** in the problem.

SOLVING A FIRST DEGREE EQUATION WITH ONE VARIABLE

- **Eliminate any fractions** by using the Multiplication Property of Equality (could be tricky if not handled properly); *EXAMPLE:* 1/2 (3a + 5) = 2/3 (7a - 5) + 9 would be multiplied on both sides of the = sign by the lowest common denominator of 1/2 and 2/3, which is 6; the result would be 3(3a + 5) = 4(7a - 5) + 54, notice that only 1/2, 2/3, and 9 were multiplied by 6 and not the contents of the parentheses; the parentheses will be handled in the next step, distribution.
- **Distribute** to remove any parentheses, if there are any. *EXAMPLE:* 3(3a + 5) = 4(7a - 5) + 54 becomes 9a + 15 = 28a - 20 + 54.
- **Combine** any like terms that are on the same side of the equals sign. *EXAMPLE:* 9a + 15 = 28a - 20 + 54 becomes 9a + 15 = 28a + 34 because the only like terms on the same side of the equals sign were the -20 and the +54.
- **Use the addition property of equality** to add the same terms on both sides of the equals sign. This may be done more than once. The objective here is to get all terms with the same variable on one side of the equals sign and all constants without the variable on the other side of the equals sign. *EXAMPLE:* 9a + 15 = 28a + 34 becomes 9a + 15 - **28a - 15** = 28a + 34 - **28a - 15**. Notice that both -28a and -15 were added to both sides of the equals sign at the same time. This results in -19a = 19 after like terms are added or subtracted.
- **Use the multiplication property of equality** to make the coefficient of the variable a 1; *EXAMPLE:* -19a = 19 would be multiplied on both sides by -1/19 (or divided by -19) to get a 1 in front of the a so the equation would become 1a = 19(-1/19) or simply a = -1.
- **Check the answer** by substituting it for the variable in the original equation to see if it works.

SOLVING A FIRST DEGREE INEQUALITY WITH ONE VARIABLE

- Follow the same steps for solving a first degree equality as described above, except for one step in the process. This exception follows.
- **Exception:** When applying the multiplication property, **the inequality sign must turn around if you multiplied by a negative number.** *EXAMPLES:* In 4m > -48, you need to multiply both sides of the > symbol by 1/4. Therefore, 4m(1/4) > -48(1/4). This results in m > -12. Notice the > did **not** turn around because you multiplied by a positive 1/4. However, in -5x > 65 you need to multiply both sides by -1/5. Therefore, -5x(-1/5) < 65(-1/5). This results in x < -13. Notice the > **did** turn around and become < because you multiplied by a negative number, -1/5.
- **Check the solution** by substituting some numerical values of the variable in the original inequality.

U.S.$4.95 / CAN.$7.50 June 2002

ISBN 157222506-8

50495

9 781572 225060

T3-ADK-553

BarCharts, Inc. ® WORLD'S #1 ACADEMIC OUTLINE

Quick Study. ACADEMIC

MATH REVIEW

multiplication.ratio.proportions.algebra

A COMPREHENSIVE OVERVIEW OF BASIC MATHEMATICAL CONCEPTS

SET THEORY

NOTATION

- **{ }** *Braces* indicate the beginning and end of a set notation; when listed, elements or members must be separated by commas; *EXAMPLE:* in A = {4, 8, 16} the 4, 8, and 16 are called elements or members of the set; sets are finite (ending, or having a last element) unless otherwise indicated.
- **. . .** In the middle of a set indicates *continuation of a pattern*; *EXAMPLE:* B = {5, 10, 15, ... , 85, 90}.
- **. . .** At the end of a set indicates an *infinite set*, that is, a set with no last element; *EXAMPLE:* C = {3, 6, 9, 12, ...}.
- **|** Is a symbol which literally means "such that"
- **∈** Means *is a member of* or *is an element of*; *EXAMPLE:* if A = {4, 8, 12} then 12 ∈ A because 12 is in set A.
- **∉** Means *is not a member of* or *is not an element of*; *EXAMPLE:* If B = {2, 4, 6, 8} then 3 ∉ B because 3 is not in set B.
- **∅** *Empty set* or *null set*: a set containing no elements or members, but which is a subset of all sets; also written as { }.
- **⊂** Means *is a subset of*; also may be written as ⊆.
- **⊄** Means *is not a subset of*; also may be written as ⊈.
- **A⊂B** Indicates that every element of set A is *also an element of* set B; *EXAMPLE:* If A = {3, 6} and B = {1, 3, 5, 6, 7, 9} then A ⊂ B because the 3 and 6 which are in set A are also in set B.
- **2^n** Is the *number of subsets of a set* when *n* equals the number of elements in that set; *EXAMPLE:* If A = {4, 5, 6} then set A has 8 subsets because A has 3 elements and 2^3 = 8.

OPERATIONS

A∪B Indicates the *union* of set A with set B; every element of this set is EITHER an element of set A, OR an element of set B; that is, to form the union of two sets, put all of the elements of the two sets together into one set, making sure not to write any element more than once; *EXAMPLE:* If A = {2, 4, 6, 8, 10, 12} and B = {3, 6, 9, 12, 15, 18} then: A ∪ B = {2, 3, 4, 6, 8, 9, 10, 12, 15, 18}.

∪ UNION

2 4 6 3 9

8 10 12 15 18

A B

A∩B Indicates the *intersection* of set A with set B; every element of this set is also an element of set A AND set B; that is, to form the intersection of two sets list only those elements which are found in BOTH of the two sets; *EXAMPLE:* If A = {2, 4, 6, 8, 10, 12} and B = {3, 6, 9, 12, 15, 18} then A ∩ B = {6, 12}.

∩ INTERSECTION

2 4 6 3 9

8 10 12 15 18

A B

$\overline{A}$ Indicates the *complement* of set A; that is, all elements in the universal set which are NOT in set A; *EXAMPLE:* If the Universal set is the set of Integers and A = {0, 1, 2, 3, ...} then $\overline{A}$ = {-1, -2, -3, -4,...}.

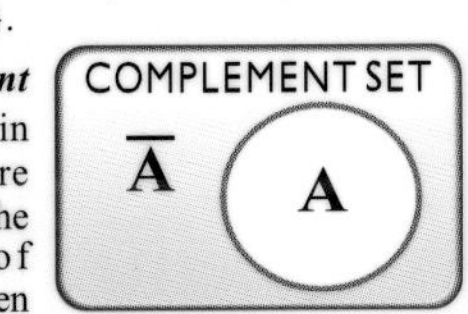

PROPERTIES

- **A = B** If all of the elements in set A are also in set B and all elements in set B are also in set A, although they do not have to be in the same order. *EXAMPLE:* If A = {5, 10} and B = {10, 5} then A = B.
- **n(A)** Indicates the *number of elements in set A*, i.e., the **cardinal number** of the set. *EXAMPLE:* If A = {2, 4, 6} then n (A) = 3.
- **A ~ B** Means *is equivalent to*; that is, set A and set B have the same number of elements, although the elements themselves may or may not be the same. *EXAMPLE:* If A = {2, 4, 6} and B = {6, 12, 18} then A ~ B because n (A) = 3 and n (B) = 3.
- **A∩B = ∅** Indicates *disjoint sets* which have no elements in common. *EXAMPLE:* if A = {3, 4, 5} and B = {7, 8, 9} then A∩B = ∅ because there are no common elements.

6 54614 20506 3

PROPERTIES OF REAL NUMBERS

CLOSURE

- **a + b** is a real number; when you add 2 real numbers, the result is also a real number. *EXAMPLE:* 3 and 5 are both real numbers, 3 + 5 = 8 and the sum, 8, is also a real number.
- **a – b** is a real number; when you subtract 2 real numbers the result is also a real number. *EXAMPLE:* 4 and 11 are both real numbers, 4 – 11 = –7, and the difference, –7, is also a real number.
- **(a)(b)** is a real number; when you multiply 2 real numbers, the result is also a real number. *EXAMPLE:* 10 and –3 are both real numbers, (10)(-3) = –30, and the product, –30, is also a real number.
- **a / b** is a real number when b ≠ 0; when you divide 2 real numbers, the result is also a real number unless the denominator (divisor) is zero. *EXAMPLE:* –20 and 5 are both real numbers, -20 / 5 = – 4, and the quotient, – 4, is also a real number.

COMMUTATIVE

- **a + b = b + a**; you can add numbers in either order and get the same answer. *EXAMPLE:* 9 + 15 = 24 and 15 + 9 = 24 so 9 + 15 = 15 + 9.
- **(a)(b) = (b)(a)**; you can multiply numbers in either order and get the same answer. *EXAMPLE:* (4)(26) = 104 and (26)(4) = 104 so (4)(26) = (26)(4).
- **a – b ≠ b – a**; you **cannot subtract in any order** and get the same answer. *EXAMPLE:* 8 – 2 = 6, but 2 – 8 = –6. There is no commutative property for subtraction.
- **a/b ≠ b/a**; you **cannot divide in any order** and get the same answer. *EXAMPLE:* 8/2 = 4 , but 2/8 = .25 so there is no commutative property for division.

ASSOCIATIVE

- **(a + b) + c = a + (b + c)**; you can group numbers in any arrangement when adding and get the same answer. *EXAMPLE:* (2 + 5) + 9 = 7 + 9 = 16 and 2 + (5 + 9) = 2 + 14 = 16 so (2 + 5) + 9 = 2 + (5 + 9).
- **(ab)c = a (bc)**; you can group numbers in any arrangement when multiplying and get the same answer. *EXAMPLE:* (4x5)8 = (20)8 = 160 and 4(5x8) = 4(40) = 160 so (4x5)8 = 4(5x8).
- The associative property *does not work* for subtraction or division. *EXAMPLES:* (10 – 4) – 2 = 6 – 2 = **4**, but 10 – (4 – 2) = 10 – 2 = **8** for division (12/6)/2 = (2)/2 = **1**, but 12/(6/2) = 12/3 = **4**. Notice that these answers are not the same.

IDENTITIES

- **a + 0 = a**; zero is the identity for addition because adding zero does not change the original number. *EXAMPLE:* 9 + 0 = 9 and 0 + 9 = 9.
- **a (1) = a**; one is the identity for multiplication because multiplying by one does not change the original number. *EXAMPLE:* 23 (1) = 23 and (1) 23 = 23.
- Identities for subtraction and division become a problem. It is true that 45 – 0 = 45, but 0 – 45 = –45, not 45. This is also the case for division because 4/1 = 4 , but 1/4 = .25, so the identities do not hold when the numbers are reversed.

INVERSES

- **a + (–a) = 0**; a number plus its additive inverse (the number with the opposite sign) will always equal **zero**. *EXAMPLE:* 5 + (–5) = 0 and (–5) + 5 = 0. The exception is zero because 0 + 0 = 0 already.
- **a (1/a) = 1**; a number times its multiplicative inverse or reciprocal (the number written as a fraction and flipped) will always equal **one**. *EXAMPLE:* 5(1/5) = 1. The exception is zero because zero cannot be multiplied by any number and result in a product of one.

DISTRIBUTIVE PROPERTY

- **a(b + c) = ab + ac or a(b - c) = ab – ac**; each term in the parentheses must be multiplied by the term in front of the parentheses. *EXAMPLE:* 4(5 + 7) = 4(5) + 4(7) = 20 + 28 = 48. This is a simple example and the distributive property is not required to find the answer. When the problem involves a variable however, the distributive property is a necessity. *EXAMPLE:* 4(5a + 7) = 4(5a) + 4(7) = 20a + 28.

PROPERTIES OF EQUALITY

- **REFLEXIVE : a = a**; both sides of the equation are identical. *EXAMPLE:* 5 + k = 5 + k.
- **SYMMETRIC : If a = b then b = a**. This property allows you to exchange the two sides of an equation. *EXAMPLE:* 4a – 7 = 9 – 7a+15 becomes 9 – 7a + 15 = 4a – 7.
- **TRANSITIVE : If a = b and b = c then a = c**. This property allows you to connect statements which are each equal to the same common statement. *EXAMPLE:* 5a – 6 = 9k and 9k = a + 2; you can eliminate the common term 9k and connect the following into one equation: 5a – 6 = a + 2.
- **ADDITION PROPERTY OF EQUALITY: If a = b then a + c = b + c.** This property allows you to add any number or algebraic term to any equation as long as you add it to **both** sides to keep the equation true. *EXAMPLE:* 5 = 5; if you add 3 to one side and not the other the equation becomes 8 = 5 which is false, but if you add 3 to both sides you get a true equation 8 = 8. Also, 5a + 4 = 14 becomes 5a + 4 + (–4) = 14 + (–4) if you add – 4 to both sides. This results in the equation 5a = 10.
- **MULTIPLICATION PROPERTY OF EQUALITY: If a = b then ac = bc when c ≠ 0.** This property allows you to multiply both sides of an equation by any nonzero value. *EXAMPLE:* If 4a = –24, then (4a)(.25) = (–24)(.25) and; a = –6. Notice that both sides of the = were multiplied by .25.

SETS OF NUMBERS

DEFINITIONS

- **NATURAL or Counting NUMBERS:** {1, 2, 3, 4, 5, ..., 11, 12, ...}
- **WHOLE NUMBERS:** {0, 1, 2, 3,..., 10, 11, 12, 13, ...}
- **INTEGERS:** {... , -4, -3, -2, -1, 0, 1, 2, 3, 4, ...}
- **RATIONAL NUMBERS: {p/q | p** and q are integers, q≠0}; the sets of Natural numbers, Whole numbers, and Integers, as well as numbers which can be written as proper or improper fractions, are all subsets of the set of Rational Numbers.
- **IRRATIONAL NUMBERS: {x|** x is a real number but is not a Rational number}; the sets of Rational numbers and Irrational numbers have no elements in common and are therefore disjoint sets.
- **REAL NUMBERS: {x |** x is the coordinate of a point on a number line}; the union of the set of **Rational numbers** with the set of **Irrational numbers** equals the set of Real Numbers.
- **IMAGINARY NUMBERS: {ai |** a is a real number and *i* is the number whose square is -1}; i^2 = -1; the sets of Real numbers and Imaginary numbers have no elements in common and are therefore **disjoint** sets.
- **COMPLEX NUMBERS: {a + bi |** a and b are real numbers and *i* is the number whose square is -1}; the set of **Real numbers** and the set of **Imaginary numbers** are both subsets of the set of Complex numbers. *EXAMPLES:* 4 + 7*i* and 3 - 2*i* are complex numbers.

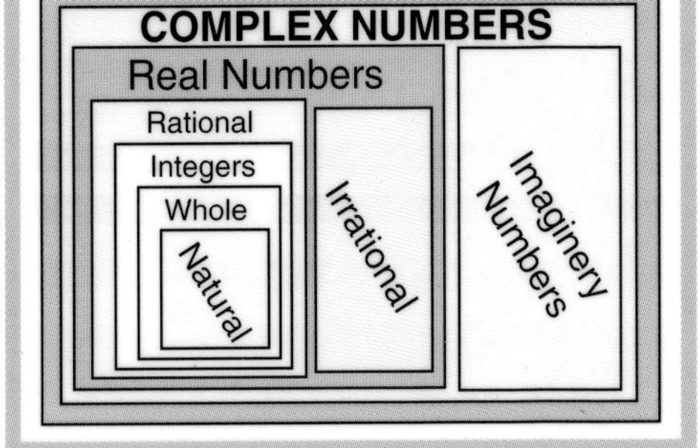

OPERATIONS OF REAL NUMBERS

VOCABULARY

- **TOTAL** or **SUM** is the answer to an addition problem. The numbers added are called **addends**. *EXAMPLE:* In 5 + 9 = 14, 5 and 9 are addends and 14 is the total or sum.
- **DIFFERENCE** is the answer to a subtraction problem. The number subtracted is called the **subtrahend**. The number from which the subtrahend is subtracted is called the **minuend**. *EXAMPLE:* In 25 - 8 = 17, 25 is the minuend, 8 is the subtrahend, and 17 is the difference.
- **PRODUCT** is the answer to a multiplication problem. The numbers multiplied are each called a **factor**. *EXAMPLE:* In 15 x 6 = 90, 15 and 6 are factors and 90 is product.
- **QUOTIENT** is the answer to a division problem. The number being divided is called the **dividend**. The number that you are dividing by is called the **divisor**. If there is a number remaining after the division process has been completed, that number is called the **remainder**. *EXAMPLE:* In 45 ÷ 5 = 9, which may also be written as 5)45 or 45/5, 45 is the dividend, 5 is the divisor and 9 is the quotient.
- An **EXPONENT** indicates the number of times the **base** is multiplied by itself; that is, used as a factor. *EXAMPLE:* In 5^3, 5 is the base and 3 is the exponent, or power, and 5^3 = (5)(5)(5) = 125, notice that the base, 5, was multiplied by itself 3 times.
- **PRIME NUMBERS** are natural numbers greater than 1 having exactly two factors, itself and one. *EXAMPLES:* 7 is prime because the only two natural numbers that multiply to equal 7 are 7 and 1; 13 is prime because the only two natural numbers that multiply to equal 13 are 13 and 1.
- **COMPOSITE NUMBERS** are natural numbers that have more than two factors. *EXAMPLES:* 15 is a composite number because 1, 3, 5, and 15 all multiply in some combination to equal 15; 9 is composite because 1, 3, and 9 all multiply in some combination to equal 9.
- The **GREATEST COMMON FACTOR (GCF)** or **greatest common divisor (GCD)** of a set of numbers is the largest natural number that is a factor of each of the numbers in the set; that is, the largest natural number that will divide into all of the numbers in the set without leaving a remainder. *EXAMPLE:* The greatest common factor (GCF) of 12, 30 and 42 is 6 because 6 divides evenly into 12, 30, and 42 without leaving remainders.
- The **LEAST COMMON MULTIPLE (LCM)** of a set of numbers is the smallest natural number that can be divided (without remainders) by each of the numbers in the set. *EXAMPLE:* The least common multiple of 2, 3, and 4 is 12 because although 2, 3, and 4 divide evenly into many numbers including 48, 36, 24, and 12, the smallest is 12.
- The **DENOMINATOR** of a fraction is the number in the bottom; that is, the divisor of the indicated division of the fraction. *EXAMPLE:* In 5/8, 8 is the denominator and also the divisor in the indicated division.
- The **NUMERATOR** of a fraction is the number in the top; that is, the dividend of indicated division of the fraction. *EXAMPLE:* In 3/4, 3 is the numerator and also the dividend in the indicated division.

FUNDAMENTAL THEOREM OF ARITHMETIC

The Fundamental Theorm of Arithmetic states that every composite number can be expressed as a unique product of prime numbers. *EXAMPLES:* 15 = (3)(5), where 15 is composite and both 3 and 5 are prime; 72 = (2)(2)(2)(3)(3), where 72 is composite and both 2 and 3 are prime, notice that 72 also equals (8)(9), but this does not demonstrate the theorem because neither 8 nor 9 are prime numbers.

ORDER OF OPERATIONS

- **DESCRIPTION:** The order in which addition, subtraction, multiplication, and division are performed determines the answer.
- **ORDER**
  1. **Parentheses:** Any operations contained in parentheses are done **first**, if there are any. This also applies to these enclosure symbols { } and [ ].
  2. **Exponents:** Exponent expressions are simplified **second**, if there are any.
  3. **Multiplication and Division:** These operations are done next in the order in which they are found, going **left** to **right**; that is, if division comes first going left to right, then it is done first.
  4. **Addition and Subtraction:** These operations are done next in the order in which they are found going **left** to **right**; that is, if subtraction comes first, going left to right, then it is done first.

DECIMAL NUMBERS

- The **PLACE VALUE** of each digit in a base ten number is determined by its position with respect to the decimal point. Each position represents multiplication by a power of ten. *EXAMPLE:* In 324, 3 means 300 because it is 3 times 10^2 (10^2 = 100). 2 means 20 because it is 2 times 10^1 (10^1 = 10), and 4 means 4 times one because it is 4 times 10^0 (10^0 = 1). There is an invisible decimal point to the right of the 4. In 5.82, 5 means 5 times one because it is 5 times 10^0 (10^0 = 1), 8 means 8 times one tenth because it is 8 times 10^{-1} (10^{-1} = .1 = 1/10), and 2 means 2 times one hundredth because it is 2 times 10^{-2} (10^{-2} = .01 = 1/100).

PLACE VALUE

WRITING DECIMAL NUMBERS AS FRACTIONS

- Write the digits that are behind the decimal point as the numerator (top) of the fraction.
- Write the place value of the last digit as the denominator (bottom) of the fraction. Any digits in front of the decimal point are whole numbers.
  EXAMPLE: In 4.068, the last digit behind the decimal point is 8 and it is in the 1000ths place; therefore, 4.068 becomes $4\frac{68}{1000}$.
- Notice the number of zeros in the denominator is equal to the number of digits behind the decimal point in the original number.

ADDITION

- Write the decimal numbers in a vertical form with the **decimal points lined up** one under the other, so digits of equal place value are under each other.
- **ADD**
  EXAMPLE: 23.045 + 7.5 + **143** + .034 would become

```
  23.045
   7.5
 143.0     because there is an invisible decimal point
    .034   behind the 143.
 -------
 173.579
```

SUBTRACTION

- Write the decimal numbers in a vertical form with the **decimal points lined up** one under the other.
- Write additional zeros after the last digit behind the decimal point in the minuend (number on top) if needed (both the minuend and the subtrahend must have an equal number of digits behind the decimal point).
- *EXAMPLE:* In 340.06 – 27.3057, 340.06 only has 2 digits behind the decimal point, so it needs 2 more zeros because 27.3057 has 4 digits behind the decimal point; therefore, the problem becomes:

```
  340.0600
 – 27.3057
```

MULTIPLICATION

- **Multiply**
- Count the number of digits behind the decimal points in all factors.
- Count the number of digits behind the decimal point in the answer. The answer must have the same number of digits behind the decimal point as there are digits behind the decimal points in all the factors. It is not necessary to line the decimal points up in multiplication. *EXAMPLE:* In (3.05)(.007), multiply the numbers and count the **5 digits** behind the decimal points in the **problem** so you can put 5 digits behind the decimal point in the **product** (answer); therefore, (3.05)(.007) = .02135 This process works because .3 times .2 can be written as fractions, 3/10 times 2/10, which equals 6/100 which equals .06 as a decimal number — two digits behind the decimal points in the problem and two digits behind the decimal point in the answer.

OPERATIONS OF INTEGERS

ABSOLUTE VALUE

- ***Definition:*** |x| = x if x > 0 or x = 0 and |x| = –x if x < 0; that is, the absolute value of a number is always the positive value of that number. ***EXAMPLES:*** |6| = 6 and |–6| = 6, the answer is positive 6 in both cases.

ADDITION

- **If the signs of the numbers are the *same*, ADD.** The answer has the same sign as the numbers.
  EXAMPLES: (–4) + (–9) = –13 and 5 + 11 = 16.
- **If the signs of the numbers are *different*, SUBTRACT.** The answer has the sign of the larger number (ignoring the signs or taking the absolute value of the numbers to determine the larger number).
  EXAMPLES: (–4)+(9) = 5 and (4)+(–9) = –5.

SUBTRACTION

- **Change subtraction to addition of the opposite number**; a - b = a + (-b); that is, change the subtraction sign to addition and also change the sign of the number directly behind the subtraction sign to the opposite. Then follow the addition rules above.
  EXAMPLES: (8) – (12) = (8) + (–12) = – 4 and (–8) – (12) = (–8) + (–12) = –20 and (–8) – (–12) = (–8) + (12) = 4. Notice the sign of the number in front of the subtraction sign never changes.

MULTIPLICATION AND DIVISION

Multiply or divide, then follow these rules to determine the sign of the answer.

- If the numbers have the **same signs the answer is *POSITIVE***.
- If the numbers have **different signs the answer is *NEGATIVE***.
- It makes no difference which number is larger when you are trying to determine the sign of the answer.
  EXAMPLES: (–2)(–5) = 10 and (–7)(3) = –21 and (–2)(9) = –18.

DOUBLE NEGATIVE

- **–(–a) = a** that is, the sign in front of the parentheses changes the sign of the contents of the parentheses. *EXAMPLES:* –(–3) = +3 or –(3) = –3; also, –(5a – 6) = –5a + 6.

DIVISION

- **Rule**: Always divide by a whole number.
- If the divisor is a whole number simply divide and bring the decimal point up into the quotient (answer).

  EXAMPLE: $4\overline{)\,.16}$ = .04

- If the divisor is a decimal number, move the decimal point behind the last digit and move the decimal point in the dividend the same number of places. Divide and bring the decimal point up into the quotient (answer).

  EXAMPLE: $.05\overline{)\,3.50}$ = 70.

- This process works because both the divisor and the dividend are actually multiplied by a power of ten, that is 10, 100, 1000, or 10000 to move the decimal point.

  EXAMPLE: $\frac{3.5}{.05} \times \frac{100}{100} = \frac{350}{5} = 70$

FRACTIONS

REDUCING

- Divide numerator (top) and denominator (bottom) by the same number, thereby renaming it to an equivalent fraction in lower terms. This process may be repeated.

  EXAMPLE: $\frac{20 \div 4}{32 \div 4} = \frac{5}{8}$

ADDITION

$\frac{a}{c} + \frac{b}{c} = \frac{a+b}{c}$ where c ≠ 0

- Change to equivalent fractions with **common denominator**. *EXAMPLE:* To evaluate $\frac{2}{3} + \frac{1}{4} + \frac{5}{6}$ follow these steps:
  1. Find the **least common denominator** by determining the smallest number which can be divided evenly (no remainders) by all of the numbers in the denominators (bottoms). *EXAMPLE:* 3, 4, and 6 divide evenly into 12.
  2. Multiply the **numerator** and **denominator** of each fraction so the fraction value has not changed but the common denominator has been obtained. *EXAMPLE:*
     $\frac{2\times4}{3\times4} + \frac{1\times3}{4\times3} + \frac{5\times2}{6\times2} = \frac{8}{12} + \frac{3}{12} + \frac{10}{12}$
  3. Add the **numerators** and keep the same **denominator** because the addition of fractions is counting equal parts.
     EXAMPLE: $\frac{8}{12} + \frac{3}{12} + \frac{10}{12} = \frac{21}{12} = 1\frac{9}{12} = 1\frac{3}{4}$

SUBTRACTION

$\frac{a}{c} - \frac{b}{c} = \frac{a-b}{c}$ where c ≠ 0

- Change to equivalent fractions with a **common denominator**.
  1. Find the **least common denominator** by determining the smallest number which can be divided evenly by all of the numbers in the denominators (bottoms).
     EXAMPLE: $\frac{7}{9} - \frac{1}{3}$
  2. Multiply the **numerator** and **denominator** by the same number so the fraction value has not changed, but the common denominator has been obtained.
     EXAMPLE: $\frac{7}{9} - \frac{1\times3}{3\times3} = \frac{7}{9} - \frac{3}{9}$
  3. Subtract the **numerators** and keep the same **denominator** because subtraction of fractions is finding the difference between equal parts.
     EXAMPLE: $\frac{7}{9} - \frac{3}{9} = \frac{4}{9}$

MULTIPLICATION

$\frac{a}{c} \times \frac{b}{d} = \frac{a\times b}{c\times d}$ where c ≠ 0 and d ≠ 0

- **Common denominators are NOT needed.**
  1. **Multiply** the numerators (tops) and multiply the denominators (bottoms) **then reduce** the answer to lowest terms.
     EXAMPLE: $\frac{2}{3} \times \frac{6}{12} = \frac{12}{36} \div \frac{12}{12} = \frac{1}{3}$
  2. **OR - reduce** any numerator (top) with any denominator (bottom) and **then multiply** the numerators and multiply the denominators.
     EXAMPLE: $\frac{\cancel{2}^1}{\cancel{3}_1} \times \frac{\cancel{6}^2}{\cancel{12}_6} = \frac{1}{3}$

DIVISION

$\frac{a}{c} \div \frac{b}{d} = \frac{a}{c} \times \frac{d}{b} = \frac{a\times d}{c\times b}$ where c≠0; d≠0; b≠0

- **Common denominators are NOT needed.**
  1. **Change division to multiplication by the reciprocal**; that is, flip the fraction in back of the division sign and change the division sign to a multiplication sign.
     EXAMPLE: $\frac{4}{9} \div \frac{2}{3}$ becomes $\frac{4}{9} \times \frac{3}{2}$
  2. Now follow the steps for multiplication of fractions as indicated above.
     EXAMPLE: $\frac{\cancel{4}^2}{\cancel{9}_3} \times \frac{\cancel{3}^1}{\cancel{2}_1} = \frac{2}{3}$

MIXED NUMBERS & IMPROPER FRACTIONS

GENERAL COMMENTS

- ***Description of mixed numbers:*** Whole numbers followed by fractions; that is, a whole number added to a fraction.
  EXAMPLE: $4\frac{1}{2}$ means $4 + \frac{1}{2}$
- **Improper fractions** are fractions that have a numerator (top number) larger than the denominator (bottom number).
- **Conversions**
  1. **Mixed number to improper fraction**: Multiply the denominator (bottom) by the whole number and add the numerator (top) to find the numerator of the improper fraction. The denominator of the improper fraction is the same as the denominator in the mixed number.
     EXAMPLE: $5\frac{2}{3} = \frac{3\times5+2}{3} = \frac{17}{3}$
  2. **Improper fraction to mixed number**: Divide the denominator into the numerator and write the remainder over the divisor (the divisor is the same number as the denominator in the improper fraction).
     EXAMPLE: $\frac{17}{5}$ means $5\overline{)\,17}$ = $3\frac{2}{5}$ (15, remainder 2)

ADDITION

- Add the whole numbers.
- Add the fractions by following the steps for addition of fractions in the fraction section of this study guide.
- If the answer has an improper fraction, change it to a mixed number and add the resulting whole number to the whole number in the answer.
  EXAMPLE: $4\frac{3}{5} + 7\frac{4}{5} = 11\frac{7}{5} = 11 + 1\frac{2}{5} = 12\frac{2}{5}$

SUBTRACTION

- **SUBTRACT THE FRACTIONS FIRST.**
  1. If the fraction of the larger number is larger than the fraction of the smaller number, then follow the steps of subtracting fractions in the fraction section of this study guide and then subtract the whole numbers.
     EXAMPLE: $7\frac{5}{6} - 2\frac{1}{6} = 5\frac{4}{6} = 5\frac{2}{3}$
  2. If that is not the case, then borrow ONE from the whole number and add it to the fraction (must have common denominators) before subtracting.
     EXAMPLE: $6\frac{2}{7} = 5 + \frac{7}{7} + \frac{2}{7} = 5\frac{9}{7}$; $-3\frac{5}{7} = -3\frac{5}{7}$; $= 2\frac{4}{7}$
- **SHORT CUT FOR BORROWING:** Reduce the whole number by one, replace the numerator by the sum (add) of the numerator and denominator of the fraction and keep the same denominator.
  EXAMPLE: $6\frac{2}{7} = 5 + \frac{2+7}{7} = 5\frac{9}{7}$; $-3\frac{5}{7} = -3\frac{5}{7}$; $= 2\frac{4}{7}$

MULTIPLICATION AND DIVISION

Change each mixed number to an improper fraction and follow the steps for multiplying and dividing fractions.

RATIO, PROPORTION, & PERCENT

RATIO

- ***Definition***: Comparison between two quantities.
- Forms: 3 to 5, 3 : 5, 3/5, $\frac{3}{5}$

PROPORTION

- ***Definition***: Statement of equality between two ratios or fractions.
- Forms: 3 is to 5 as 9 is to 15, 3 : 5 :: 9 : 15, $\frac{3}{5} = \frac{9}{15}$

PERCENTS

- ***Definition***: **Percent** means "out of 100" or "per 100."
- **Percents and equivalent fractions**
  1. Percents can be written as fractions by placing the number over 100 and simplifying or reducing.
     EXAMPLES: $30\% = \frac{30}{100} = \frac{3}{10}$; $4.5\% = \frac{4.5}{100} = \frac{45}{1000} = \frac{9}{200}$
  2. Fractions can be changed to percents by writing them with denominators of 100. The numerator is then the percent number. *EXAMPLE:* $\frac{3}{5} = \frac{3\times20}{5\times20} = \frac{60}{100} = 60\%$
- **Percents and decimal numbers**
  1. To change **a percent to a decimal number, move the decimal point 2 places to the left** because percent means "out of 100" and decimal numbers with two digits behind the decimal point also mean "out of 100." *EXAMPLE:*
     45% = .45 ; 125% = 1.25 ; 6% = .06 ; 3.5% = .035
     because the 5 was already behind the decimal point and is not counted as one of the digits in the "move two places"
  2. To change a decimal number to a percent, **move the decimal point two places to the right.** *EXAMPLES:*
     .47 = 47%; 3.2 = 320%; .205 = 20.5%

SOLVING PROPORTIONS

- Change the fractions to equivalent fractions with **common denominators**, set numerators (tops) equal to each other, and solve the resulting statement. *EXAMPLES:*
  $\frac{3}{4} = \frac{n}{20}$ becomes $\frac{15}{20} = \frac{n}{20}$, so n = 15 ;
  $\frac{n+3}{7} = \frac{10}{14}$ becomes $\frac{n+3}{7} = \frac{5}{7}$, so n + 3 = 5 and n = 2
- **Cross multiply** and solve the resulting equation. **NOTE: cross multiplication is used to solve <u>proportions only</u> and may NOT be used in fraction multiplication.** Cross multiplication may be described as the **product** of the means being equal to **the product of the extremes.** *EXAMPLES:*
  $\frac{n}{7} = \frac{3}{5}$, 5n = 21, n = 21 ÷ 5, n = $4\frac{1}{5}$
  $\frac{3}{4} = \frac{7}{n+2}$, 3n + 6 = 28, 3n = 22, n = $7\frac{1}{3}$

GEOMETRIC FORMULAS

PERIMETER: The perimeter, P, of a two-dimensional shape is the sum of all side lengths.
AREA: The area, A, of a two-dimensional shape is the number of square units that can be put in the region enclosed by the sides. NOTE: Area is obtained through some combination of multiplying heights and bases, which always form 90º angles with each other, except in circles.
VOLUME: The volume, V of a three-dimensional shape is the number of cubic units that can be put in the region enclosed by all the sides.

Square Area: A = hb
If h=8 then b=8 also, as all sides are equal in a square, then:
A=64 square units

Rectangular Prism Volume
V=lwh; If l=12
w=3, h=4 then:
V=(12)(3)(4), V=144 cubic units

Rectangle Area A=hb, or A=lw
If h=4 and b=12
then: A = (4)(12)
A=48 square units

Cube Volume, $V=e^3$
Each edge length, e, is equal to the other edges in a cube.
If e=8 then: V=(8)(8)(8), V=512 cubic units

Triangle Area: A=1/2 bh
If h=8 and b=12 then:
A=1/2 (8)(12)
A=48 square units

Cylinder Volume, $V=\pi r^2h$
If radius, r=9, h=8 then:
$V=\pi(9)^2(8)$, V=3.14(81)(8),
V=2034.72 cubic units

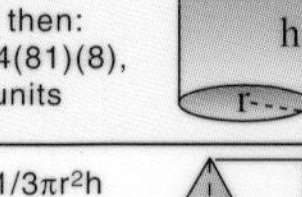

ParallelogramArea: A=hb;
If h=6 and b=9
then: A=(6)(9)
A=54 square units

Cone Volume, $V=1/3\pi r^2h$
If r=6 and h=8 then:
$V=1/3\pi(6)^2(8)$,
V=1/3(3.14)(36)(8)
V=301.44 cubic units

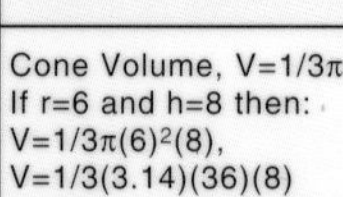
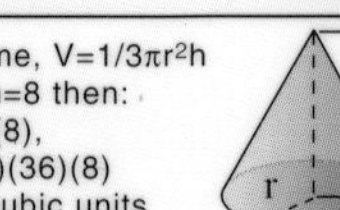

Trapezoid Area: $A=\frac{1}{2}h(b_1+b_2)$
If h=9 and b_1=8 and b_2=12
then: $A=\frac{1}{2}(9)(8+12)$
$A=\frac{1}{2}(9)(20)$
A=90 square units

Triangular Prism Volume,
V=(area of triangle)h
If [triangle], has an area equal to $\frac{1}{2}$(5)(12) then:
V=30h and if h=8 then:
V=(30)(8), V=240 cubic units

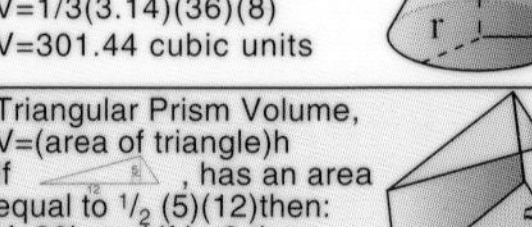

Circle Area: $A=\pi^2$
If π=3.14 and r=5
then: $A=(3.14)(5)^2$,
A=78.5 square units
Circumference: C=2πr
C=(2)(3.14)(5)=31.4 units

Rectangular Pyramid Volume
$V=\frac{1}{3}$(area of rectangle)h
If 1 = 5 and w = 4
the rectangle has an area of 20, then:
$V=\frac{1}{3}$(20)h and if h=9
then: $V=\frac{1}{3}$(20)(9), V=60c.u.

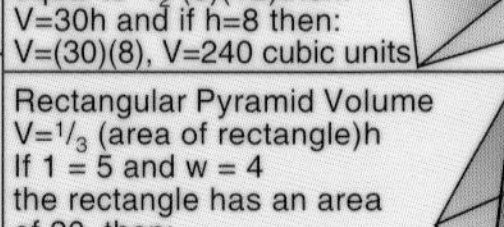
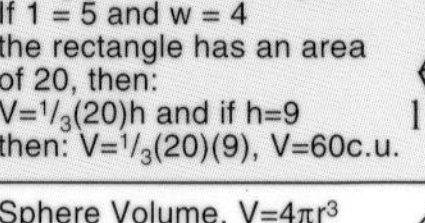

PYTHAGOREAN THEOREM
If a right triangle has hypotenuse c and sides a and b,
then $c^2=a^2+b^2$

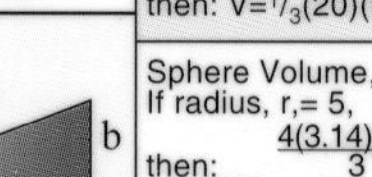
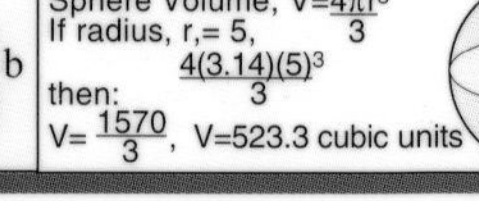

Sphere Volume, $V=\frac{4\pi r^3}{3}$
If radius, r,= 5,
then: $\frac{4(3.14)(5)^3}{3}$
$V=\frac{1570}{3}$, V=523.3 cubic units

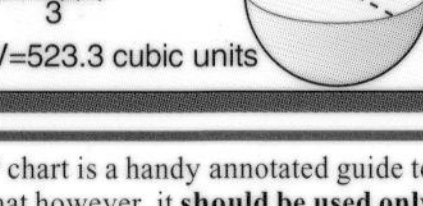
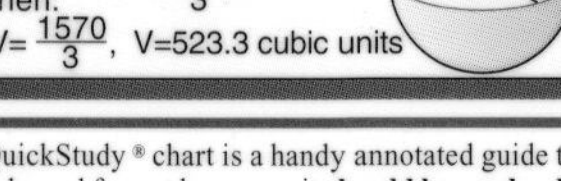